YOUR KNOWLEDGE HAS VALUE

- We will publish your bachelor's and master's thesis, essays and papers

- Your own eBook and book - sold worldwide in all relevant shops

- Earn money with each sale

Upload your text at www.GRIN.com and publish for free

Imprint:

Copyright © 2018 GRIN Verlag
Print and binding: Books on Demand GmbH, Norderstedt Germany
ISBN: 9783668746640

This book at GRIN:

https://www.grin.com/document/430846

Zhen Li Ang

The Void Hypothesis. An Alternate Gravity Hypothesis to Explain Several Gravitational Anomalies

GRIN Verlag

VOID HYPOTHESIS

Written by Ang Zhen Li
Graduate Student of Nanyang Technological University with Bachelor in Science of Physics

ABSTRACT

This is an Alternate Gravity Hypothesis used to explain several gravitational anomalies that occur on galactic scales.

Content Page

Premise:

Newton's Law of Gravity is wrong.

Hypothesis:

The fundamental force of gravity is a result of space repelling space, matter occupying space in 3-dimensional space experiences attraction wherein the space between 2 separate masses are compressed, and repulsion wherein the space between 2 separate masses are stretched. The Hypothesis asserts that space is elastic and can therefore be compressed or stretched. The Hypothesis asserts that a unit space exerts a repulsive force on another unit space universally.

Symbols Defined

F_G	Gravitational Force calculated by Newton's Law of Gravity
F_{void}	Space-Repulsion Force calculated by Void Hypothesis
V_{in}	Volume of inner enclosed region of space
V_{out}	Volume of outer enclosed region of space
C_{uni}	Universal Constant of Space-Repulsion
V_1	Volume of empty space in enclosed region 1
V_2	Volume of empty space in enclosed region 2
$M_\odot$	Solar Mass
V_0	Volume occupied by a Hydrogen Atom at lowest Energy level
R_{earth}	Mean radius of the earth
R_{venus}	Mean distance between planetary orbits of Venus and Earth
R_U	User set upper limit of Space-Repulsion
g_{void}	Gravitational field strength of Space-Repulsion
$g_{Newtonian}$	Gravitational field strength of Newtonian-Gravity
A_{volume}	Volume of region A
B_{volume}	Volume of region B
C_{volume}	Volume of region C
L_{volume}	Volume of region L
M_{volume}	Volume of region M
R_{volume}	Volume of region R
$P1_{resultant}$	Resultant vector of unoccupied unit space at point P1
$P1_{compression}$	Compression magnitude of unoccupied unit space at point P1
$P2_{resultant}$	Resultant vector of unoccupied unit space at point P2
$P2_{compression}$	Compression magnitude of unoccupied unit space at point P2

Introduction

The premise is vital in order to explain the hypothesis which theorises that the fundamental force of Gravity is not a result of matter exerting an attraction force on matter but a result of space repelling space. In this scientific paper, the Void Hypothesis shall be used to explain the following anomalies that occur on galactic scales in the Universe:

(i) Galaxies have less mass than is observed otherwise known as the 'missing mass problem'. (Oakes, 2010)
(ii) Stars further from the centre of a galaxy are not orbiting slower past a critical point. (Matthews, 2018)
(iii) Rotation speeds of Dwarf Galaxies are higher than that of Galaxies. (Brownstein & Moffat, 2005)
(iv) Gas from the giant spherically shaped Galaxy – M87 did not disperse after a long time. (NASA)
(v) Tully-Fisher Relation: The amount of light energy emitted by a spiral galaxy is roughly proportional to its speed of rotation. The faster the galaxies spin, the brighter they are. (Matthews, 2018)
(vi) Dark Flow: Many large galaxy clusters converging to an area in the night sky between constellations Vela and Centaurus. (Gefter, 2009)

Many of these anomalies can be explained directly by that the gravitational force a galaxy exerts on itself is greater in magnitude according to Void Hypothesis than Newtonian-Gravity.

Void Hypothesis Formulation

Each unoccupied unit space in the Universe repels every other unoccupied unit space (m^3) at a universal constant which results in expansion or compression of regions of space. Consider that masses that occupy space are at fixed Cartesian coordinates in space and does not move along the fabric of space, but rather the regions of space within and around and between them stretches, and in so doing moves the masses. Hence, when two discrete masses are converging, the region of space in between the masses is compressing; conversely when two discrete masses are diverging, the region of space in between the masses is expanding. According to this definition of gravity, the Space-Repulsion Force exerted to compress or expand a defined volume of space is directly proportional to the product of the two regions of space.

$F_{void} \propto V_{in} V_{out}$, V_{in} and V_{out} are the respectively defined and separate enclosed regions of space repelling each other. $F_{void} = C_{uni} V_{in} V_{out}$, C_{uni} is the universal constant by which two unoccupied unit spaces repel each other.

Setting Parameters

Since the force generated by unoccupied unit spaces is universal and undiminished by distance, it would be difficult to understand without first defining enclosed regions of space in which space beyond this enclosure does not generate forces influencing the space within the enclosure.

There is a general distinction between masses in a large region of space and masses in a small region of space. For a small enclosed space with isolated masses concentrating near the centre, the masses converge towards each other because the space within the enclosure that is surrounding the isolated masses repels the space between the isolated masses. This is comparable to the effect of Newtonian-Gravity wherein asteroids, meteorites or comets that fall within the dominant gravitational field of a star, planet, dwarf planet or moon is pulled towards it. In contrast, for a large space scale the space between the isolated masses could be so large that its space-repulsion within could cause it to expand rather than compress. Hence, isolated masses or mass systems along the coordinates of radial expansion will experience divergence resulting in large mass systems like galaxies and galactic clusters to be moving apart.

Convergence of Masses

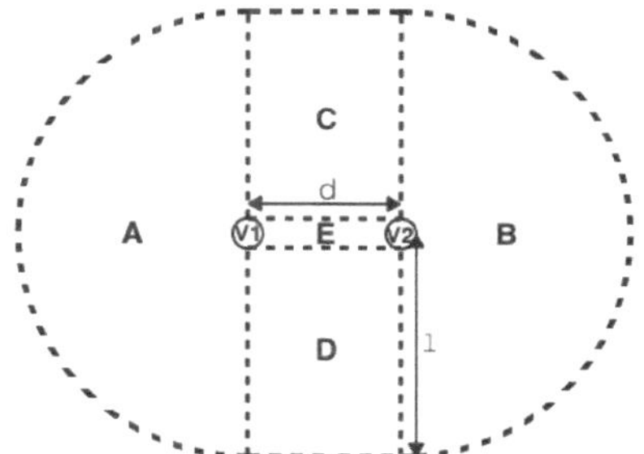

Figure 1: Cross-section of a 3-dimensional pill-shaped region of space with two discrete and identical spherical masses of uniform mass density occupying volumes V1 and V2 respectively.

The two discrete masses are not point masses and by conventional physics is experiencing Newtonian-Gravity attraction towards each other without orbiting each other. The distance between the centres of the two masses, d is small compared to the radius of the hemispherical regions A and B, l. (d << l); the enclosed volume E is much smaller than enclosed volumes A, B, C and D. (E << {A, B, C, D}); E is being repelled from all directions by the surrounding A, B, C and D and is compressed. As E is being compressed by F_{void}, the two masses being of fixed coordinates on the fabric of space accelerates towards each other. Each respective enclosed volume of regions of space must also repel itself from within and is expanding outward at its own respective F_{void}. However, E is expanding outward at a much smaller F_{void} than the overall compressional forces F_{void} exerted on E by A, B, C and D. The enclosed volume V1 and V2 also contain empty space within the solid lattice matter of the masses. However, considering V1 and V2 is much smaller than E, the forces F_{void} generated by this space on A, B, C, D and E is small and insignificant.

Alternative to considering bulk regions of space consider the vector forces of unoccupied unit spaces at coordinates within the region of spaces A, B, C and D.

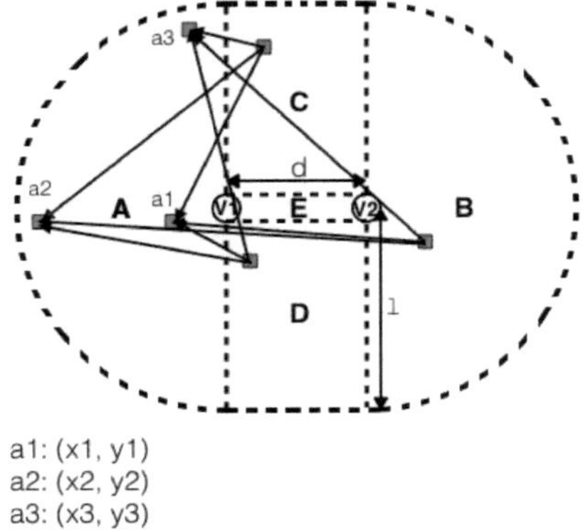

a1: (x1, y1)
a2: (x2, y2)
a3: (x3, y3)

Figure 2: : The vector space-repulsion forces acting on three defined points in the region A by each arbitrary point in regions B, C and D. (Considering rotational symmetry of pill shape, exclude z axis.)

Consider the vector components acting on an unoccupied unit space that negates each other are added together as the magnitude of compression of the unoccupied unit space; Consider the vector components acting on an unoccupied unit space that add to each other are added together to form a resultant vector with a magnitude and direction. If the magnitude of compression exceeds the magnitude of resultant vector, the unoccupied unit space is compressed. If the magnitude of resultant vector exceeds the magnitude of compression, the unoccupied unit space is stretched in the direction of the resultant vector by the difference between the two magnitudes.

B, C and D each have an arbitrary point pointing towards a1, a2 and a3 in A. a1 and a3 each have a greater resultant vector than a2. a2 have a greater compression magnitude than a1 and a3.

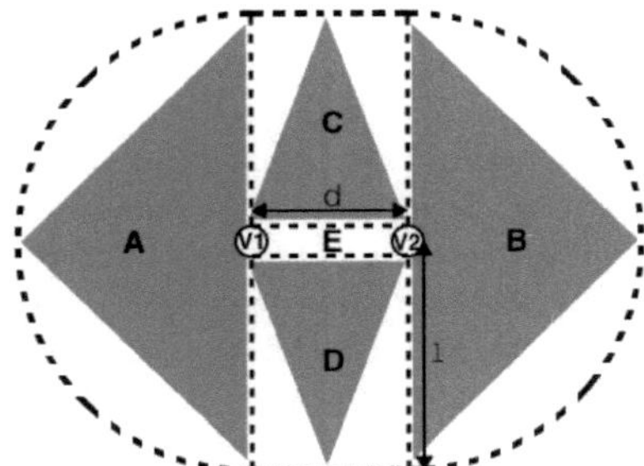

Figure 3: The cones are used to illustrate spatially the result of the vector space repulsion forces.

Hence, one can draw a cone with a curved bottom to accommodate V1 in A, and associate the points approaching the slant height of the cone from the inside as regions of increasing resultant vectors and decreasing compression magnitudes. Conversely associate points approaching the base of the cone with the curved inset around V1 as regions of increasing compression magnitude and decreasing resultant vectors. Do the same for B, C and D.

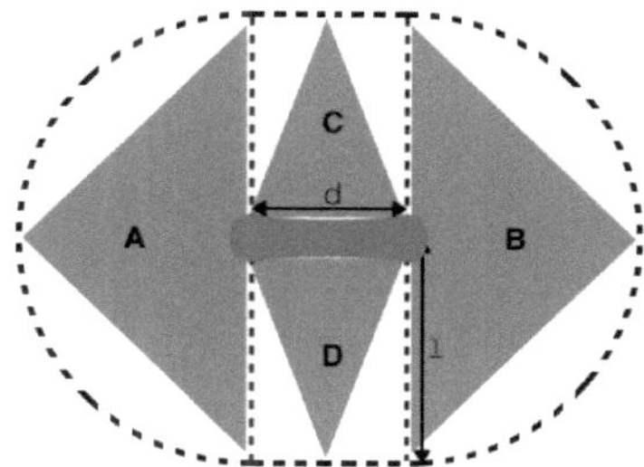

Figure 4: Smooth out the edges of the cones and join the inner parts of the cones.

In consideration of edge effects smooth out the edges of the cones and draw an imaginary line to join up and form the region enclosed by the cones (blue) as a region of high compression magnitude and low resultant vectors. This blue region seems reminiscent of what a gravitational field of the two isolated masses would look like in Newtonian-Gravity.

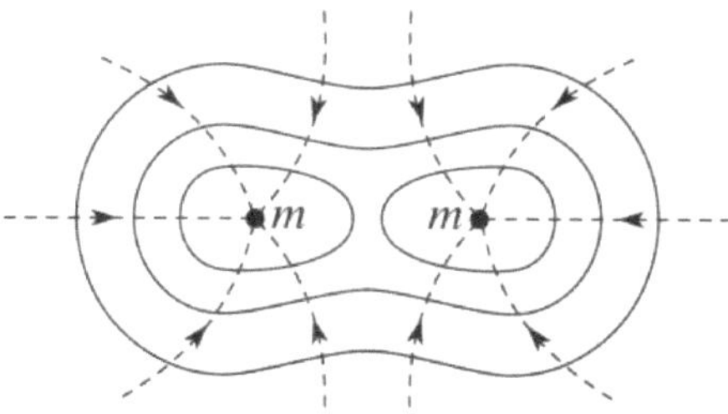

Figure 5: Equipotential contours (solid curves) surrounding two masses, m close together. Gravitational field lines (dashed lines).

Divergence of Masses

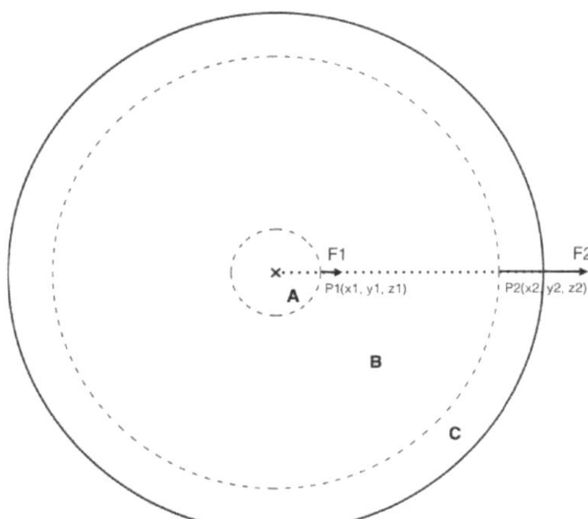

Figure 6: Cross-section of 3-dimensional space of the Universe (assumed spherical) with the enclosed regions of volume of spaces C surrounding B, B surrounding A.

The centre of the Universe is marked by an X and the edge of the Universe is characterised by a Bold Circle. The two points in space P1 and P2 are respectively a point near the centre of the Universe, and a point near the edge of the Universe.

A is much smaller than B and C combined, and the F_{void} exerting on the surface area of A by B and C is the product of A_{volume} and the addition of B_{volume} and C_{volume}.

$$F_{void} = C_{uni} A_{area} (B_{area} + C_{area})$$

C is much smaller than A and B combined and the F_{void} exerting on the inner surface area of C by A and B is the product of C_{volume} and the addition of A_{volume} and B_{volume}.

$$F_{void} = C_{uni} C_{area} (A_{area} \quad B_{area})$$

Calculations will show that the F_{void} acting on P1 is much smaller than the F_{void} acting on P2. Assign the radius of A, B and C at ratio of 1, 10, 11 units respectively.

$$A_{volume} = \frac{4}{3}\pi(1^3) = \frac{4}{3}\pi \text{ (No units)}$$

$$B_{volume} = \frac{4}{3}(10^3) - \frac{4}{3}\pi(1^3) = \frac{4}{3}\pi(999) \text{ (No units)}$$

$$C_{volume} = \frac{4}{3}(11^3) - \frac{4}{3}\pi(10^3) = \frac{4}{3}\pi(331) \text{ (No units)}$$

$$B_{volume} + C_{volume} = \frac{4}{3}\pi(1330) \text{ (No units)}$$

$$A_{volume} + B_{volume} = \frac{4}{3}\pi(1000) \text{ (No units)}$$

$$A_{volume}(B_{volume} + C_{volume}) = \frac{4}{3}\pi \times \frac{4}{3}\pi(1330) = (\frac{4}{3}\pi)^2(1330) \text{ (No units)}$$

$$C_{volume}(A_{volume} + B_{volume}) = \frac{4}{3}\pi(331) \times \frac{4}{3}\pi(1000) = (\frac{4}{3}\pi)^2(331000) \text{ (No units)}$$

A mass system at P1 is accelerating radially outwards at a slower rate than a mass system at P2. Since the Universe is expanding outwards at an accelerating rate both vectors should be starting from their respective points P1 and P2, and pointing right of the Figure 2. (Gefter, 2009)

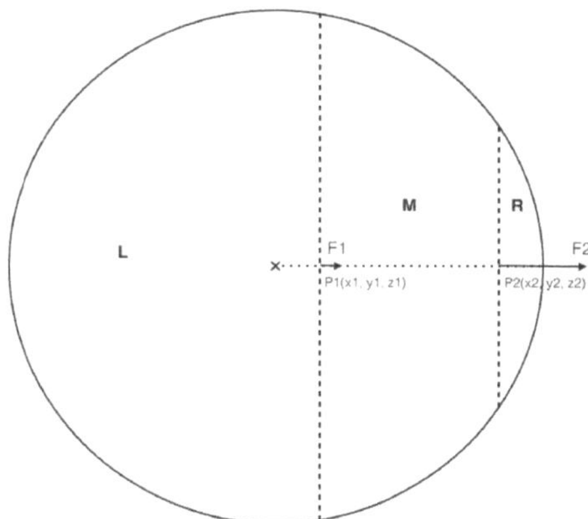

Figure 7: Cross-section of 3-dimensional space of the Universe (assumed spherical) with the enclosed regions of volume of spaces L, M and R.

Each unit space in L, M and R exerts a vector of equal magnitude and different directions on the unit space at P1. Due to symmetry of the spherical Universe, the resultant vector from L on P1 and P2 is to the right and any resultant from M and R on P1 and P2 is to the left, but since L is greater than M and R combined, the resultant of all the unit spaces in the Universe on the unit space at P1 is to the right; likewise since L and M combined is greater than R, the resultant of all the unit spaces in the Universe on the unit space at P2 is also to the right. The compression magnitude on P1 and P2 would be a result of all vectors components perpendicular to the horizontal acting on the two respective unoccupied unit spaces. Hence, separate the L, M and R regions horizontally across the mid-section and calculate the effects of the perpendicular components on the points P1 and P2.
Using previously assigned ratios for A, B and C; distance from the centre of the universe to P1, P2 and the edge of the Universe is respectively 1, 10 and 11 units.
Calculations will show that F2 is much greater than F1, and the compression magnitude at P1 is much greater than at P2.

Volume of a spherical cap: $V = \dfrac{\pi h^2}{3}(3r - h)$

$$L_{volume} = \frac{4}{3}\pi(11^3) - \frac{\pi 10^2}{3}(3(11) - 10) = \frac{\pi}{3}(4(1331) - 2300) = \frac{\pi}{3}(3024) \text{ (No units)}$$

$$M_{volume} = \frac{\pi 10^2}{3}(3(11) - 10) - \frac{\pi 1^2}{3}(3(11) - 1) = \frac{\pi}{3}(2300 - 32) = \frac{\pi}{3}(2268) \text{ (No units)}$$

$$R_{volume} = \frac{\pi 1^2}{3}(3(11)-1) = \frac{\pi}{3}(32) \text{ (No units)}$$

$$\left|F1_{resultant}\right| : L_{volume} - M_{volume} - R_{volume} = \frac{\pi}{3}(3024 - 2268 - 32) = \frac{\pi}{3}(724) \text{ (No units)}$$

$$P1_{compression} = (\frac{L_{volume}}{2})^2 (\frac{M_{volume} + R_{volume}}{2})^2$$

$$= (\frac{\pi}{3})^2 (\frac{724}{2})^2 (\frac{2268 + 32}{2})^2 = (\frac{\pi}{3})^2 (362^2)(1150^2) = (\frac{\pi}{3})^2 (2^2 \cdot 5^2 \cdot 23 \cdot 181)^2 \text{ (No units)}$$

$$\cong 1.9 \times 10^{11}$$

$$\left|F2_{resultant}\right| = L_{volume} + M_{volume} - R_{volume} = \frac{\pi}{3}(3024 + 2268 - 32) = \frac{\pi}{3}(5260) \text{ (No units)}$$

$$P2_{compression} = (\frac{L_{volume} + M_{volume}}{2})^2 (\frac{R_{volume}}{2})^2$$

$$= (\frac{\pi}{3})^2 (\frac{3024 + 2268}{2})^2 (\frac{32}{2})^2 = (\frac{\pi}{3})^2 (2646^2)(16^2) = (\frac{\pi}{3})^2 (2^5 \cdot 3^2 \cdot 7^2)^2 \text{ (No units)}$$

$$\cong 2.0 \times 10^9$$

Since an unoccupied unit space at P1 has lower resultant vector and greater compression magnitude than that at P2 a mass system at P1 should be more compact and moving away from the centre of the Universe at a slower rate than an identical mass system at P2.

Hence, the same problem can be solved in terms of overall simplified area product of enclosed volumes of spaces and in terms of vectors of unit spaces.

Mass and Distance Scales

Since Void Hypothesis is postulated to be universal, it should be consistently applicable at small or large scales. Consider a mass system existing in space. Depending on how large this mass system is there is an average distance between this mass system and another mass system of approximately the same order of magnitude. The effect of the space beyond this average distance should be neglected for the sake of setting an upper limit at this distance because the C_{uni} is not known. Hence, the various mass scales are defined in order to differentiate between the scales of the F_{void} acting on the mass systems.

(1) Microscopic with average distance between atoms
(2) Stars and Planets
(3) Galaxies at Intergalactic distances
(4) Galactic Clusters

Impossible to determine C_{uni}

An experimental C_{uni} for some mass scales may be calculated and used to estimate the final value of C_{uni}. Each Mass Scale is enclosed by a larger space spherical shell. For instance galaxies are enclosed by a space spherical shell of intergalactic distance thickness. As the volume of space enclosed increases for larger Mass Scales the experimental C_{uni} decreases. In the Universe, the larger the Mass Scale the more masses would appear on different coordinates in the domain of spherical shell which would increase the complexity of the enclosed volume. Hence, the error of calculation would increase, and at the largest Mass Scale galactic clusters are surrounded by a spherical shell as large as the universe. This undiminished C_{uni} cannot be calculated with sufficient precision and accuracy because of lack of information of the unobservable Universe.

Graphical Analysis

Using Data in Graphing

Investigate the relationship between Space-Repulsion Force over Universal Constant of Space-Repulsion, F_{void}/C_{uni} against the Radial Distance from the Centre of Earth, r.

Acquire data points at the surface of each Earth layer {Crust, Upper Mantle, Lower Mantle, Outer Core, Inner Core} in addition to the Earth's centre.

Plot two graphs of F_{void}/C_{uni} against r for $0 \leq r \leq R_{earth}$ and $0 \leq r \leq R_{venus}$ respectively; where R_{earth} is the mean radius of the Earth and R_{venus} is the mean distance from Venus to Earth's orbit.

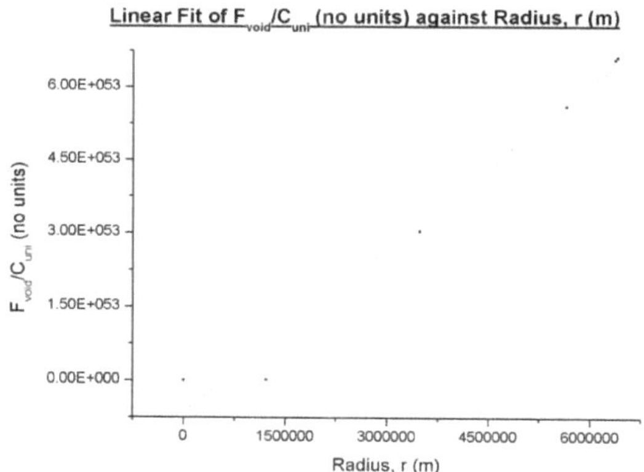

Figure 8: Linear Fit of data points within Earth's Radius

The linear fit of data points for $0 \leq r \leq R_{Earth}$ shows that the graph varies linearly with a positive gradient.

Generate 100 data points with equally spaced divisions between R_{earth} and R_{venus}.

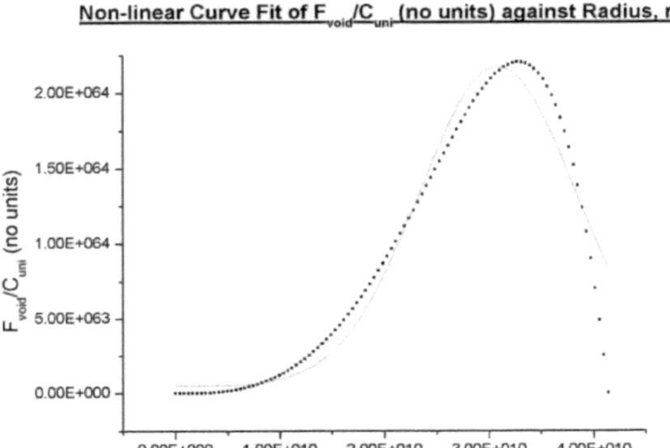

Figure 9: Non-linear Fit of data points until Venus's orbit

The exponential non-linear curve fit of data points for $0 \leq r \leq R_{venus}$ shows that the graph seems to resemble the graph of equation: $y = x^3(R_U^3 - x^3)$ where $R_U = R_{venus}$.

This graph shows that the gravitational force increases as r increases beyond the Earth's surface. This shows that the implementation of Space-Repulsion on Mass Scale (2) is erroneous. This failure in representation can be attributed to one factor. The upper limit of this graph, R is erroneously speculated to be at R_{venus} because the C_{uni} is unknown.

Attempting to use Space-Repulsion to represent the Mass Scale of (3) or (4) may also be erroneous because the planets and stars are part of a greater mass system the galaxy; each planetary and star sized object generates its own gravitational field according to the equation: $y = x^3(R_U^3 - x^3)$ with respect to its centre and with its own respective R value, and increases the complexity of finding the resultant graph for the galaxy.

Using Arbitrary Equation and Graph

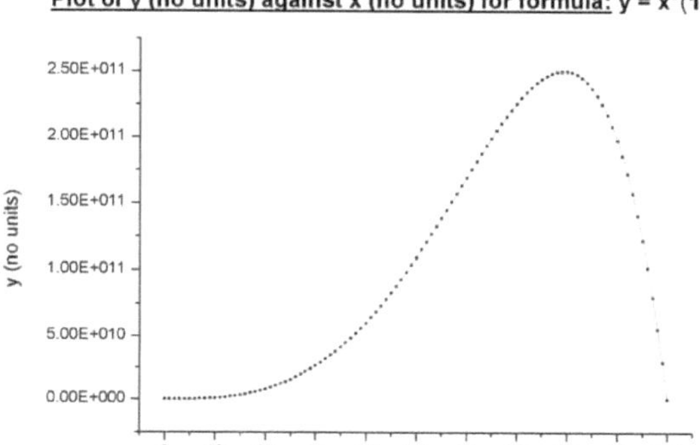

Figure 10: Graph of Arbitrary Equation

Use the Arbitrary Graph of formula $y = x^3(R_U{}^3 - x^3)$, $R_U = 100$, to explain the galaxies dimensions. y represents the gravitational field strength whereas x represents the radial distance from the centre of the mass system. There is a stationary point of relative maximum point at (79.8, 250 000 000 000) (3 s.f) which coincides with the algebraic differentiation acquired expression: $x = \dfrac{R_U}{\sqrt[3]{2}}$.

The upper limit of the edge of the galaxy should be at the end point x = 100 where gravitational field strength = 0.

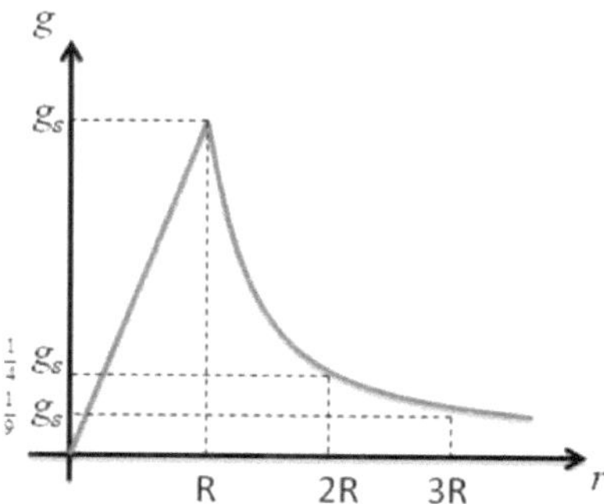

Figure 11: Typical graph of Newtonian-Gravitational Field Strength, g against Radial Distance from the Centre of a planet or star, r

$$g = \frac{F_G}{m} = -\frac{GM}{r^2}$$, where m is unit mass, M is the Mass of arbitrary planet or star and R is the radius of the planet or star.

For the condition: $0 \leq r \leq R$, M is directly proportional to r^3 because there are 3 dimensions of space while r^2 is the denominator of g. Hence, g is directly proportional to r. For the condition: $R \leq r \leq \infty$, M is a constant, while r^2 is the denominator of g. Hence, g is inversely proportional to r^2.

Assume that for a galaxy, the g increases linearly as r increases across the massive bulky centre of the galaxy then starts to increase non-linearly as r increases across the less dense outer part of the galaxy, and lastly starts to decrease at a polynomial degree of -2 beyond the edge of the galaxy. Such a graph can be surmised to look roughly like Figure 11 wherein the sharp discontinuous peak is smoothed out. Hence, consider the sharp discontinuous peak in Figure 11 to be a narrow peak for the comparison below.

Compare Figures 10 and 11 by aligning their starting, peak and end points respectively and proportionately. Figure 11 does not have a graph end point; hence simply pick a point where $g_{Newtonian}$ is extremely small. Figure 10 has a broad peak whereas Figure 11 has a narrow peak. Assume that the Space-Repulsion Field Strength peak is greater in magnitude than the Newtonian-Gravitational Field Strength peak since g_{void} increases at an exponential rate to peak whereas $g_{Newtonian}$ increases at a linear rate to peak.

Based on the above comparison, the calculations of the orbital linear velocity of stars in the less dense outer part of the galaxy using Newtonian-Gravity would yield the gravitational anomaly (ii). This means that the edge of the galaxy is somewhere beyond R in Figure 11 for astronomer's calculations.
Based on the same comparison, calculations of the orbital linear velocity of stars about the galactic centre in the less dense outer part of the galaxy using Space-Repulsion would yield cohesive results if the edge of the galaxy is somewhere beyond the stationary point or relative maximum value but extremely close to it. Because the peak is broad and the rate of change of g_{void} as r changes is small, the stars in the less dense outer part of the galaxy would seem to orbit the galactic centre at roughly the same orbital linear velocity as r increases.

Even as the limit of the galaxy is at the end point of Figure 10, the edge of the galaxy is not a well-defined distance because the galaxy is a discontinuous mass of loosely held stars and planets. Stars of low but non-zero centripetal force might either be flung out of the galaxy into intergalactic space or drawn into orbit around stellar masses deeper into the galaxy during the formation of the galaxy, thus, creating a massless band of space in-between the broad peak and the end point of the graph. Hence, the existing stars at the less dense outer part of the galaxy would have substantial centripetal force acting on it.

In circular motion, $F_{centripetal} = \dfrac{mv^2}{r}$

Newtonian-Gravity,

$$F_G = \dfrac{GMm}{r^2}$$

$$F_{centripetal} = F_G$$

$$\dfrac{mv^2}{r} = \dfrac{GMm}{r^2}$$

$$v = \sqrt{\dfrac{GM}{r}}$$

Space-Repulsion,

$$F_{void} = C_{uni}V_{in}V_{out}, V_{in} = \dfrac{4}{3}\pi r^3, V_{out} = \dfrac{4}{3}\pi R_U^3 - \dfrac{4}{3}\pi r^3$$

$$F_{void} = C_{uni}(\dfrac{4}{3}\pi r^3)(\dfrac{4}{3}\pi)(R_U^3 - r^3) = C_{uni}(\dfrac{4}{3}\pi)^2(r^3)(R_U^3 - r^3)$$

$$F_{centripetal} = F_{void}$$

$$\dfrac{mv^2}{r} = C_{uni}(\dfrac{4}{3}\pi)^2(r^3)(R_U^3 - r^3)$$

$$v = (\dfrac{4}{3}\pi r^2)\sqrt{\dfrac{C_{uni}(R_U^3 - r^3)}{m}}$$

The orbital linear velocities of stars in orbit about the galaxy's centre for Newtonian-Gravity is dependent on the mass of the galaxy, whereas for Space-Repulsion is dependent on the mass of the star and the radial distance from the galaxy's centre.

In Figure 10, there might be a point where the gravitational field strength drops to a value equal to 100 billionth of the Earth's surface gravity in-between the broad peak and the end point of the graph. Given that the mid-point of the board peak and the end point of the graph is a steep negative gradient, it may have been misunderstood as a critical-point where gravitational constant rapidly decreases. (Matthews, 2018)

Explanation of Gravitational Anomalies

(i) Galaxies have less mass than is observed otherwise known as the 'missing mass problem'. (Oakes, 2010)

Explanation:
In the Mass Scales (3) and (4), Newtonian-Gravitational Force is weaker than Space-Repulsion Force in magnitude and hence, there is a missing mass when calculating the mass of a galaxy based on Newtonian-Gravity and the observed light coming from it. Calculations using Void Hypothesis may yield an accurate value.

(ii) Stars further from the centre of a galaxy are not orbiting slower past a critical point. (Matthews, 2018)

Explanation:
According to Newtonian-Gravity, stars further from the centre of a galaxy should have a lower Gravitational Force acting on it and hence orbit the galaxy centre at a lower linear speed. This is because of a direct proportionality relationship between the centripetal force of a circular motion system and the orbital linear velocity of the object in motion ($F \propto v^2$). However, according to Space-Repulsion, the Graphical Analysis assumes that the maximum peak is broader and is at a larger magnitude than that in Newtonian-Gravity. Assuming that the less dense outer part of the galaxy fits the broad peak the orbital linear velocity of stars would remain roughly the same since there is only a small change in centripetal force as radial distance from the centre of the galaxy increases.

(iii) Rotation speeds of Dwarf Galaxies are higher than that of Galaxies. (Brownstein & Moffat, 2005)

Dividing the predicted MSTG flat rotation velocity over the MSTG predicted total mass of the galaxy and then averaging for the following categories:

Dwarf (LSB & HSB) Galaxies: 85.5 ($\dfrac{kms^{-1}}{10^{10} M_\odot}$) (1 d.p)

LSB Galaxies: 43.7 ($\dfrac{kms^{-1}}{10^{10} M_\odot}$) (1 d.p)

HSB Galaxies: 27.8 ($\dfrac{kms^{-1}}{10^{10} M_\odot}$) (1 d.p) (Brownstein & Moffat, 2005)

Explanation:
Dwarf Galaxies are of lower mass and size than Galaxies. Assume that intergalactic distance applies for Dwarf Galaxies; a Dwarf Galaxy has a smaller inner volume, V_{in} and larger outer volume, V_{out} as compared to a Galaxy. According to Space-Repulsion, and the calculations involving Figure 3a, the centripetal force of a Dwarf Galaxy acting on itself is higher than that of a Galaxy acting on itself as compared to Newtonian-Gravity, and as before ($F \propto v^2$). Hence, the stars of a Dwarf Galaxy rotate about its centre faster than those of a Galaxy.

The error in calculating the orbital velocity is further amplified by the Newtonian-Gravitational Force having a direct proportionality relationship with the mass of the mass system ($F \propto M$).

(iv) Gas from the giant spherically shaped Galaxy – M87 did not disperse after a long time. (NASA)

Explanation:
The gas cloud of the giant spherically shaped Galaxy has not yet dispersed because the Space-Repulsion Force acting on a Galaxy sized gas cloud is greater than that calculated in Newtonian-Gravity. Hence, the gas cloud of the Galaxy M87 is still visible even as it is only supposed to exist for an approximate 100 million years. (NASA)

(v) Tully-Fisher Relation: The amount of light energy emitted by a spiral galaxy is roughly proportional to its speed of rotation. The faster the galaxies spin, the brighter they are. (Matthews, 2018)

Explanation:
The spiral arms of a galaxy have enclosed volumes of empty space between each spiral arm and these spaces are not symmetrical with respect to a spiral arm. The empty space that are in front and behind each spiral arm acts with Space-Repulsion Forces on the empty space between the stellar components of the spiral arm and disturbs the respective velocities of the stars and planets. Perhaps this causes them to move unpredictably by a small extent outside of their intended trajectory with respect to the galaxy. The stars which were originally obscured by other stars blocking in front of it comes out behind the star and into view of an observer. Since there is much more empty space than mass in a galaxy the probability of a star being obscured by another star before and then being revealed after is much higher than the probability of a star not being obscured by another star before and being obscured by it afterwards. Therefore, the overall effect of this disruption is that the spiral arms appear to be brighter; coincidentally the spiral arms might be orbiting faster as a result of the aforementioned Space-Repulsion Forces.

(vi) Dark Flow: Many large galaxy clusters converging to an area in the night sky between constellations Vela and Centaurus. (Gefter, 2009)

Explanation:
The Universe has been expanding uniformly in space since the Big Bang. A sudden convergence of galactic clusters is paradoxical to the expansion of the Universe. Assume the converging galactic clusters are at a part of the Universe where it has more matter distribution than other parts, and thus have less space between the galactic clusters. The Universe expands until a point where the space surrounding all the aforementioned galactic clusters is large enough that the Space-Repulsion Force that compresses the space between the galactic clusters is overwhelms the Universe's Space Repulsion Force to expand the space between the galactic clusters. Hence, the galactic clusters converge rather than diverge.

Internet Sourced Data

Earth Crust's Composition (Earth's Composition and Structure: A Journey to the Center of the Earth)

Element	% Weight	Atomic Mass
O	46.6	15.999
Si	27.7	28.085
Al	8.1	26.982
Fe	5.0	55.845
Ca	3.6	40.078
Na	2.8	22.99
K	2.6	39.098
Mg	2.1	24.305
--	1.5	

Earth's Mantle (Allegre, Poirier, Humler, & Hofmann, 1994)

Element	% Weight	Atomic Mass
O	44.79	15.999
Si	21.521	28.085
Al	2.164	26.982
Mg	22.784	24.305
Fe	5.818	55.845
Ca	2.308	40.078
K	0.028	39.098
Na	0.264	22.99
Mn	0.116	54.938
Ti	0.112	47.867
Ni	0.2	58.693
Cr	0.27	51.996

Earth's Core (Allegre, Poirier, Humler, & Hofmann, 1994)

Element	% Weight	Atomic Mass
Fe	79.39 ± 2	55.845
Ni	4.87 ± 0.3	58.693
Si	7.35	28.085
S	2.3 ± 0.2	32.06
O	4.1 ± 0.5	15.999

Earth's Layers (Structure of the Earth: HyperPhysics)

	Thickness, t (km)	Density, ρ (g/cm^3)
Crust	30	2.2
Upper Mantle	720	3.4
Lower Mantle	2171	4.4
Outer Core	2259	9.9
Inner Core	1221	12.8

Planetary Orbits (Elert, 1998)

Planet	Perihelion (10^6 km)	Aphelion (10^6 km)
Venus	107.48	108.94
Earth	147.48	152.1
Mars	206.92	249.23

Scientific Constants Used

(CODATA 2006 based on a least squares adjustment of data from different measurements
Numbers in parentheses represents the uncertainties of the last two digits)
Avogadro Number, N_A = 6.02214179 (30) x 10^{23} particle/mol
Bohr Radius, a_o = 5.2917720859 (36) x 10^{-11} m
Calculated Bohr Volume, V_0 = 6.207146595 x 10^{-31} m^3

Processed Data

	t (m)	ρ (g/m³)	V (m³) (6 d.p)	M_r (g/mol)	f_r (no units) (9 d.p)	$V \times f_r$ (m³) (6 d.p)
Venus to Earth	4.1385 x 10^{10}	--	2.969050 x 10^{32}	--	1	2.969050 x 10^{32}
Crust	3 x 10^4	2.2 x 10^6	1.537410 x 10^{19}	23.8$\pm$0.1	0.966639906	1.486122 x 10^{19}
Upper Mantle	7.2x10^5	3.4 x 10^6	3.426810 x 10^{20}	23.95$\pm$0.03	0.949613057	3.254141 x 10^{20}
Lower Mantle	2.171 x 10^6	4.4 x 10^6	9.220480 x 10^{20}		0.935745598	8.628022 x 10^{20}
Outer Core	2.259 x 10^6	9.9 x 10^6	1.090960 x 10^{21}	50.7$\pm$0.3	0.931913159	1.016676 x 10^{21}
Inner Core	1.221 x 10^6	1.28 x 10^7	7.624930 x 10^{18}		0.913689951	6.966818 x 10^{18}

[The calculated errors of uncertainties add up such that f_r would be round to 1. Hence, use the

highest precision that computer offers for processed data f_r which is at 9 decimal places.]

Radius, r	$\dfrac{F_{void}}{C_{uni}}$ (m^6) (6 d.p)
At Earth Centre	0
At Inner Core Surface	2.068483 x 10^{51}
At Outer Core Surface	3.039248 x 10^{53}
At Lower Mantle Surface	5.600950 x 10^{53}
At Upper Mantle Surface	6.567121 x 10^{53}
At Earth Crust Surface	6.611245 x 10^{53}
Interplanetary Surface	0

[F_{void}/C_{uni} acquired by multiplying the V_{out} and V_{in} at the corresponding interface r values. Errors of
uncertainty calculated from previous table would be too small at > 21 decimal places. Hence, User
set precision to 6 decimal places for processed data of V, $V \times f_r$ and the final data F_{void}/C_{uni}.]

Equations Used

Void Hypothesis

$$F_{void} = C_{uni} V_{in} V_{out}$$

Newton's Law of Gravity

$$F_G = -\frac{GMm}{r^2}$$

Fraction of Volume Composing of Unoccupied Space

$$f_r = \frac{\dfrac{M_r}{\rho L V_0}}{\dfrac{M_r}{\rho L V_0} + 1}, f_r < 1$$

Volume of Unoccupied Space

$$V \times f_r$$

Volume of a spherical cap

$$V = \frac{\pi h^2}{3}(3r - h)$$

Bibliography

Allegre, C. J., Poirier, J.-P., Humler, E., & Hofmann, A. W. (1994). *The chemical composition of the Earth*. Paris, Cedex: Elsevier Earth and Planetary Science Letters.

Brownstein, J. R., & Moffat, J. W. (2005). *Galaxy Rotation Curves Without Non-Baryonic Dark Matter*. Waterloo, Ontario: The Perimeter Institute for Theoretical Physics, Waterloo, Ontario, N2J 2W9, Canada.

Earth's Composition and Structure: A Journey to the Center of the Earth. (n.d.). Retrieved from App State: http://www.appstate.edu/~marshallst/GLY1101/Lectures/2-Earth-Composition_Structure.pdf

Elert, G. (1998). *Astronomical Data: The Physics Hypertextbook*. Retrieved from The Physics Hypertextbook: physics.info/astronomical

Gefter, A. (2009, January 21). *New Scientist*. Retrieved from New Scientist: https://www.newscientist.com/article/mg20126921-900-dark-flow-proof-of-another-universe/

Matthews, P. R. (2018, February). Something's Wrong with Gravity. *BBC earth Asia Edition Vol 9 Issue 12*, pp. 28 - 35.

NASA. (n.d.). *22 The Mystery of the Missing Mass*. Retrieved from SP-466 The Star Splitters: https://history.nasa.gov/SP-466/ch22.htm

Oakes, K. (2010, September 24). *basic space: MACHOs, WIMPs and the mystery of the missing mass*. Retrieved from basic space: https://kellyoakes.wordpress.com/2010/09/24/machos-wimps-and-the-mystery-of-the-missing-mass/

Ptable. (n.d.). Retrieved from Ptable: https://www.ptable.com

Structure of the Earth: HyperPhysics. (n.d.). Retrieved from HyperPhysics: Hyperphysics/phy-astr.gsu.edu/hbase/Geophys/earthstruct.html

YOUR KNOWLEDGE HAS VALUE

- We will publish your bachelor's and
  master's thesis, essays and papers

- Your own eBook and book -
  sold worldwide in all relevant shops

- Earn money with each sale

Upload your text at www.GRIN.com
and publish for free